Jack's Insects

Narration and Nature Study Notebook

by Karen Smith

Jack's Insects Narration and Nature Study Notebook
© 2012, Karen Smith

All rights reserved. No part of this work may be reproduced or distributed in any form by any means—graphic, electronic, or mechanical, including photocopying, recording, taping, or storing in information storage and retrieval systems—without written permission from the publisher.

> If you are a parent or teacher you may duplicate pages for yourself and students in your immediate household or classroom. Please do not duplicate pages for friends, relatives outside your immediate household, or other teachers' students.

Cover Design: John Shafer
Photos: Ruth Shafer

ISBN 978-1-61634-181-7 printed
ISBN 978-1-61634-182-4 electronic download

Published and printed by
Simply Charlotte Mason, LLC
PO Box 892
Grayson, Georgia 30017
United States

SimplyCharlotteMason.com

Contents

How to Use this Notebook 4
Materials Needed 4
How to Draw Insects. 6
Lesson 1: Meet Jack and Maggie 9
Lesson 2: Great Morpho Butterfly 13
Lesson 3: Great Bird-winged Butterfly 17
Lesson 4: Hidden Butterflies 21
Lesson 5: Caterpillar Actors and Artists. 25
Lesson 6: Spiders and a Praying Mantis. 29
Lesson 7: The Leaf-grasshopper and the Ants 33
Lesson 8: Locusts 35
Lesson 9: Katydid 41
Lesson 10: Katydid Concert 43
Lesson 11: Cicada 45
Lesson 12: Cicada Development 47
Lesson 13: Great Golden Digger Wasp. 51
Lesson 14: Candle Fly and Lantern Fly 55
Lesson 15: Cucujo 59
Lesson 16: Fireflies. 63
Lesson 17: Diving-Bell Spider 67
Lesson 18: Raft Spider 71
Lesson 19: Swarming Bees. 75
Lesson 20: Unusual Bees and the Honey-pot Ant 77
Exam Questions 83
Who Was William Kirby?. 85
Master Insect and Spider List 86
My Drawings of Butterflies and Moths 91
My Drawings of Caterpillars 93
My Drawings of Hopping Insects 94
My Drawings of Cicadas and Wasps 96
My Drawings of Luminous Insects. 98
My Drawings of Spiders. 99
My Drawings of Bees and Ants 101
My Drawings of Flies, Beetles, and Bugs 103

How to Use this Notebook

This book is your narration and nature notebook for *Jack's Insects*. Each lesson will walk you through a chapter of *Jack's Insects*, plus suggest other books and websites to help you learn more about the insects you will meet. You'll find plenty of space to write about what you learn (narration) and to record which insects you see in your neighborhood and beyond (nature study). Don't feel that you must complete a lesson in only one day; you may want to spread some lessons over several days. That's fine.

The suggested books were in print (unless otherwise noted) and the websites were current when this guide was written. You'll find some of the older books free to read online. If you can't find one of the books suggested, don't worry; check your local library for other books that cover the same type of insect and use those. If a website is no longer available, you can ask your parent to help you perform an Internet search for the insect you're researching.

The main thing is to enjoy reading *Jack's Insects* and see what you can learn about those fascinating insects God created!

Materials Needed

Jack's Insects

Field guides for insects and butterflies. Use what you can find. Our favorites include
  National Wildlife Federation: Field Guide to Insects and Spiders of North America by Arthur V. Evans
  Princeton Field Guides: Caterpillars of Eastern North America by David L. Wagner

Pencils, colored pencils, or watercolors for recording observations (If you want to paint pictures of the insects, get a sketch book with thick paper.)

Field Trips: Bug Hunting, Animal Tracking, Bird-watching, Shore Walking with Jim Arnosky (optional nature study reference book but highly recommended)

Lesson 2
  Insect Investigators: Entomologists (Scientists at Work) by Richard Spilsbury

Lesson 3
  The Life Cycles of Butterflies: From Egg to Maturity, a Visual Guide to 23 Common Garden Butterflies by Judy Burris and Wayne Richards
  (optional) *An Extraordinary Life: The Story of a Monarch Butterfly* by Laurence Pringle (excellent, but out of print)

Lesson 4
  Jungle Bugs: Masters of Camouflage and Mimicry by Bruce Purser (This book is used for pictures only. The text is full of evolutionary thinking.)

Lesson 6
  Praying Mantises: Hungry Insect Heroes (Insect World) by Sandra Markle

Social Life in the Insect World by Jean Henri Fabre (This book is available free at http://simplycm.com/bf/98099.)

Lesson 8
Locusts: Insects on the Move by Sandra Markle
The Fiddlehoppers: Crickets, Katydids, and Locusts by Phyllis Perry
Social Life in the Insect World by Jean Henri Fabre
The Bible
On the Banks of Plum Creek by Laura Ingalls Wilder

Lesson 10
Grasshoppers and Crickets of North America by Sara Swan Miller

Lesson 12
Cicadas!: Strange and Wonderful by Laurence Pringle
Social Life in the Insect World by Jean Henri Fabre

Lesson 13
Animal Lives: Bees and Wasps (Qeb Animal Lives) by Sally Morgan

Lesson 14
The Glow-Worm and Other Beetles by Jean Henri Fabre (This book is available free at http://simplycm.com/bf/98101.)

Lesson 16
Fireflies (Bugs Bugs Bugs) by Margaret Hall

Lesson 17
Spiders by Seymour Simon

Lesson 20
Bramble-Bees and Others by Jean Henri Fabre (This book is available free at http://simplycm.com/bf/98102.)

How to Draw Insects

1. Draw the shapes.

Most everything you draw is made up of common shapes like circles, triangles, and ovals. So when you observe an insect, look for those basic shapes and where they are in relation to each other.

2. Draw the connecting lines.

Of course, insects are not just several shapes stuck together like a snowman. The basic shapes will form a foundation that you can add to and tweak to better reflect what the insect really looks like.

3. Fill in any patterns or special markings.

If you look closely at an insect, you will notice that it is usually not a solid color. Many insects have patterns or special markings on their bodies or wings. Those markings are part of the way God made them unique and often serve a greater purpose than just decoration. Try to faithfully duplicate the patterns and markings where they belong, then you will be ready to add color.

4. Add the color.

Color plays an important part in an insect's life. It is part of their natural camouflage or mimicry. (You'll learn more about those self-defense techniques as you read *Jack's Insects*.) So try to match the correct colors as closely as possible, whether you use colored pencils or paint.

Take a look at how we followed those steps to create the insect drawing on the next page, then try your hand at following the steps with the three insects on page 8.

Helpful Reminders
- Get as close to the insect you're drawing as is safely possible. If you can safely catch the insect in a clear glass or plastic container, you might put it in the refrigerator for 15–30 minutes. The cold will slow the insect down, making it easier to observe and draw. When you are done drawing it, release it back where you originally found it.
- Try to draw each insect its actual size.
- Label your drawings.
- Remember that these drawings are not just art exercises. You are drawing the insects to help you observe, learn, and fully appreciate how God made them.
- Listen for the sounds of insects to help you know they are around and to locate them.

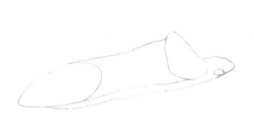

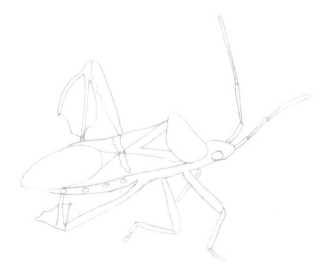
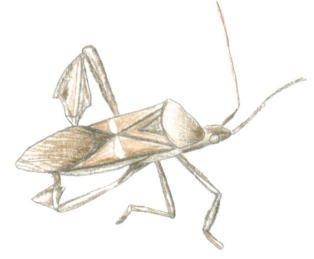

Lesson 1

Meet Jack and Maggie

Read *Jack's Insects*, chapter 1, "Inside the Book."

Narration

1. Describe Jack's view of entomology.

_____
_____
_____
_____

2. Describe Maggie's thoughts about insects and entomology.

_____
_____
_____
_____
_____

3. Tell about Jack and Maggie's conversation about books and how they got into the book.

_____
_____
_____
_____
_____

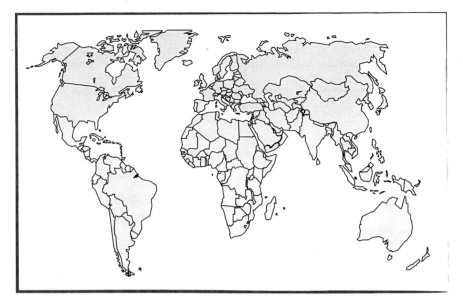

On this world map, mark and label the continent where Jack and Maggie found the first insect inside the book: the Great Morpho Butterfly.

Describe its habitat in that country.

_____
_____
_____

Find Out More

Use a dictionary to define "entomology."

_____
_____
_____
_____

Nature Study

As you read through *Jack's Insects*, plan time to go outside and look for insects in person. There will be reminders at the ends of the lessons in this book to guide you. You will also find several drawing pages in the back of this book. Drawing is a great way to force yourself to look closely and carefully.

So when you go outside, grab this notebook and flip to the back. You can keep track of which insects and spiders you have seen by checking them off the list on pages

86 through 90. And you can draw them on the labeled pages as you come across the different insects.

If it is too cold outside to find any insects, don't worry. Learn all you can about them now, then when it warms up you can go outside and use what you know to help you find and identify them. Or feel free to draw from a photograph if you can't find any insects or it is too cold to find them.

Reminder: Get Insect Investigators: Entomologists (Scientists at Work) for lesson 2.

Lesson 2

Great Morpho Butterfly

Read *Jack's Insects*, chapter 2, "Jack Promises the Great Morpho Butterfly."

Narration

1. Describe the first insect Jack and Maggie meet and how it defined *real* entomology.

_____
_____
_____
_____
_____

2. Explain the butterfly's reasons for not killing insects.

_____
_____
_____
_____
_____
_____

3. Describe the promise that was demanded of Jack.

_____
_____
_____
_____

In 1910, when Jack's Insects was written, collecting butterflies was a hobby. Many people collected the chrysalises and butterflies not only for their own collections but to sell to others. A great example of this hobby is Gene Stratton Porter's book, Girl of the Limberlost. Though some people still collect butterflies and other insects today, it is not on the scale that it was when Jack's Insects was written.

Find Out More

The discussion with the Great Morpho butterfly brings up some interesting questions to ponder. Though the butterfly presented a sometimes popular opinion, let's take some time to think it through. The book *Insect Investigators: Entomologists* (Scientists at Work) by Richard Spilsbury will give you additional information as you consider the questions below.

1. Is man's collecting insects really a great detriment to insect populations? Why?

   _____
   _____
   _____
   _____

2. What are some good reasons for killing insects?

   _____
   _____
   _____
   _____

3. What benefits are there to studying dead insects?

   _____
   _____
   _____
   _____

4. What are some other things that could cause a species to become extinct?

   _____
   _____
   _____
   _____

"But even this will not bring you to the end of your pleasures: you must leave the dead to visit the living; you must behold insects when full of life and activity, engaged in their several employments, practising their various arts, pursuing their amours, and preparing habitations for their progeny: you must notice the laying and kind of their eggs; their wonderful metamorphoses; their instincts, whether they be solitary or gregarious; and the other miracles of their history — all of which will open to you a richer mine of amusement and instruction, I speak it without hesitation, than any other department of Natural History can furnish" (William Kirby, An Introduction to Entomology, letter I).

If you would like to see a picture and read more information on Morpho butterflies, visit http://simplycm.com/jacks0201. Remember to ask your parent's permission before visiting any website.

Here's what I found out . . .

_____
_____
_____
_____
_____
_____
_____
_____
_____
_____

Nature Study

When you are outside doing nature study, see what butterflies and moths you can find in your area. You may want to keep track of which ones you have seen by checking them off the list on pages 86 and 87.

On pages 91 and 92 you can draw the butterflies and moths that you see. If you don't see them live, you can draw them by looking at a photograph. Be sure to label each drawing.

Reminder: Get The Life Cycles of Butterflies: From Egg to Maturity, a Visual Guide to 23 Common Garden Butterflies *for lesson 3. If you can find a copy of* An Extraordinary Life: The Story of a Monarch Butterfly, *get that out-of-print gem for lesson 3 also.*

Lesson 3

Great Bird-winged Butterfly

Read *Jack's Insects*, chapter 3, "A Quite Unintentional Insult."

Narration

1. Tell about the conversation between Jack and the Great Bird-winged butterfly regarding the commonness of butterflies.

_____
_____
_____
_____
_____
_____
_____

2. Tell about the unintentional insult.

_____
_____
_____
_____
_____
_____

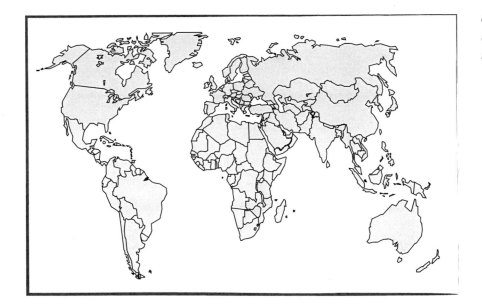

On this world map, mark and label where the Great Bird-winged Butterfly lives.

Find Out More

Use a dictionary to define "archipelago."

_____
_____
_____
_____

Read chapter 1 of *The Life Cycles of Butterflies: From Egg to Maturity, a Visual Guide to 23 Common Garden Butterflies* by Judy Burris and Wayne Richards. Use the other chapters in the book to help you find common butterflies and caterpillars.

If you can find a copy, read *An Extraordinary Life: The Story of a Monarch Butterfly* by Laurence Pringle. This book is out of print, but may be available at the library and is well worth finding. It follows the life cycle of the Monarch butterfly from egg to overwintering in Mexico. (Caution: There are a few mentions of millions of years in this book.)

If you would like to see pictures and read more information on bird-wing butterflies, you may visit these websites. Ask your parent's permission before visiting any website.

http://simplycm.com/jacks0301 This website has pictures of a bird-wing butterfly that lives in India. This is a different species from the one in *Jack's Insects* but will give you an idea of what they look like.

http://simplycm.com/jacks0302 This website has pictures and videos of several species of bird-wing butterflies. There are ads on the page for butterfly nets and books.

Here's what I found out . . .

Nature Study

When you are outside doing nature study, look for butterflies and moths.

On pages 91 and 92 you can draw the butterflies and moths that you see. Remember that these drawings are not just art exercises. You are drawing the insects to help you observe, learn, and fully appreciate how God made them.

Reminder: Get Jungle Bugs: Masters of Camouflage and Mimicry or another insect field guide for lesson 4.

Lesson 4

Hidden Butterflies

Read *Jack's Insects*, chapter 4, "A Butterfly with a Grievance And Other Butterflies."

Narration

1. Describe how the Leaf Butterfly disguised himself as a leaf.

___

2. Explain how the Leaf Butterfly used camouflage to help him survive.

___

3. Tell all you know of mimicry.

___

4. Tell about Heliconea's grievance.

_____
_____
_____
_____
_____
_____
_____

Find Out More

Use a field guide, *Jungle Bugs: Masters of Camouflage and Mimicry* by Bruce Purser, or another similar book to help you discover more insects that use camouflage or mimicry. List your findings on the next page.

Caution: *Jungle Bugs: Masters of Camouflage and Mimicry* is full of evolution, but the pictures are fantastic.

If you would like to see more pictures of camouflaged insects you may visit these websites. As always when using the Internet, ask your parent's permission before visiting any website.

http://simplycm.com/jacks0401 This website has photographs of camouflaged insects.

http://simplycm.com/jacks0402 This website has more photographs of camouflaged butterflies and moths. It has a picture of a butterfly similar to the one in *Jack's Insects* that looks like a dead leaf. It also has some information on each insect pictured.

http://simplycm.com/jacks0403 This site has pictures of heliconea butterflies. (Caution: Other pages on this web site do contain evolution.)

> "To such perfection, indeed, has nature in them carried her mimetic art, that you would declare, upon beholding some insects, that they had robbed the trees of their leaves to form for themselves artificial wings, so exactly do they resemble them in their form, substance, and vascular structure; some representing green leaves, and others those that are dry and withered. Nay, sometimes this mimicry is so exquisite, that you would mistake the whole insect for a portion

of the branching spray of a tree" (William Kirby, An Introduction to Entomology, letter I).

Here's what I found out . . .

1. Insects that use camouflage

_____
_____
_____
_____
_____
_____
_____
_____
_____
_____

2. Insects that use mimicry

_____
_____
_____
_____
_____
_____
_____
_____
_____
_____

Nature Study

When you are outside doing nature study, look for butterflies and moths. Keep track of which ones you have seen by checking them off the list on pages 86 and 87.

On pages 91 and 92 you can draw the butterflies and moths that you see. Try to draw each insect its actual size.

Lesson 5

Caterpillar Actors and Artists

Read *Jack's Insects*, chapter 5, "Caterpillars Extraordinary."

Narration

1. Describe how the caterpillar made himself look like a snake.

_____
_____
_____
_____
_____

2. Tell how the ant-caterpillar made himself look like an ant.

_____
_____
_____
_____

3. Describe how the flower-bud caterpillar disguised himself.

_____
_____
_____
_____
_____

4. Tell why the leaf caterpillar called himself an artist.

_____
_____
_____
_____
_____
_____

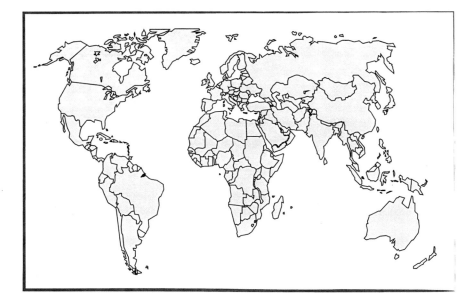

On this world map, mark and label the two locations of the flower-bud and leaf caterpillars.

Find Out More

If you would like to see pictures of caterpillars similar to those mentioned in *Jack's Insects* you may visit these websites. Remember to ask your parent's permission when using the Internet.

http://simplycm.com/jacks0501 This website has some technical information about snake caterpillars, but has some very good pictures.

http://simplycm.com/jacks0502 A naturalist's blog post and pictures about three camouflaged caterpillars.

http://simplycm.com/jacks0503 Photographs of camouflaged caterpillars.

"We may call the instincts of animals those unknown faculties implanted in their constitution by the Creator, by which, independent of instruction, observation, or experience, and without a knowledge of the end in view, they are impelled to the performance of certain actions tending to the well-being of the individual and the preservation of the species" (William Kirby, An Introduction to Entomology, letter XXVII).

Here's what I found out . . .

_____
_____
_____
_____
_____
_____
_____
_____
_____
_____
_____
_____
_____
_____
_____
_____
_____

Nature Study

When you are outside doing nature study, look for caterpillars.

On page 93 you can draw the caterpillars that you see. If you want to collect a caterpillar so you can observe firsthand its metamorphosis from caterpillar to butterfly, check out the tips on this website: http://simplycm.com/jacks0504.

Reminder: Get Praying Mantises: Hungry Insect Heroes (Insect World) *and* Social Life in the Insect World *for lesson 6.* Social Life in the Insect World *is available free at* http://simplycm.com/bf/98099.

Lesson 6

Spiders and a Praying Mantis

Read *Jack's Insects*, chapter 6, "Very Surprising Adventures."

Narration

1. Describe how Jack and Maggie were "picked."

_____
_____
_____

2. In your own words, explain the spiders' lectures to Jack and Maggie.

_____
_____
_____
_____
_____
_____
_____

3. Tell how the mantis used camouflage to catch the moth.

_____
_____
_____
_____
_____
_____

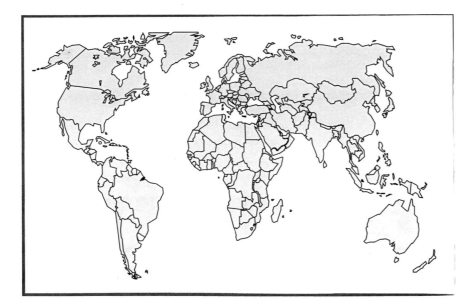

On this world map, mark and label the two places where Jack and Maggie traveled to interact with the spiders and the praying mantis. (You may need to look back at the end of chapter 5 to see where the spiders lived.)

Find Out More

Read *Praying Mantises: Hungry Insect Heroes* (Insect World) by Sandra Markle.

To find out even more about praying mantises read chapters 5 through 7, "The Mantis–The Chase," "The Mantis–Courtship," and "The Mantis–The Nest," in *Social Life in the Insect World* by Jean-Henri Fabre.

Jean-Henri Fabre was a French entomologist who lived around the time that Jack's Insects was written. Mr. Fabre made many observations of insects, including some very detailed ones of praying mantises.

This website, http://simplycm.com/jacks0601, has pictures of praying mantises that blend in with the flowers. Remember to ask your parents' permission before visiting any website. This website has ads for various products.

Here's what I found out . . .

_____
_____
_____
_____
_____
_____
_____
_____
_____
_____
_____
_____
_____
_____
_____
_____
_____
_____
_____
_____

Nature Study

When you are outside doing nature study, look for spiders and praying mantises.

On page 99 you can draw the spiders that you see, and the praying mantises on page 103. If you don't see them live, you can draw them by looking at a photograph. If you are drawing while looking at a live insect or spider, try to get as close as is safely possible so you can see more details.

Lesson 7

The Leaf-grasshopper and the Ants

Read *Jack's Insects*, chapter 7, "Jack and Maggie Have Their Lives Saved."

Narration

1. What is an "entomological eye"?

___

2. Describe the march of the ants and how the Leaf-grasshopper escaped them.

___

3. The Leaf-grasshopper called himself complex. Tell what he meant by that.

___

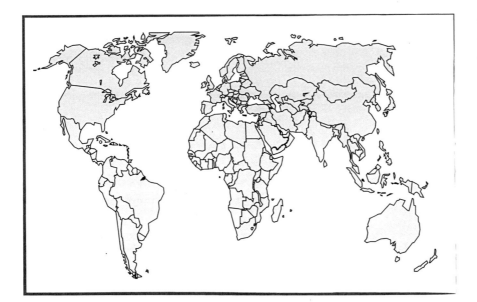

On this world map, mark and label where Jack and Maggie traveled to interact with the leaf-grasshopper and the ants. (You may need to look back at the end of chapter 6.)

Nature Study

When you are outside doing nature study, look for grasshoppers and ants. Keep track of which ones you have seen by checking them off the list on pages 88–90.

On page 94 you can draw the grasshoppers that you see, and the ants on page 101. If you don't see them live, you can draw them by looking at a photograph. You may want to use the refrigerator trick when you observe these insects. Gently catch the insect in a clear glass or plastic container and put it in the refrigerator for 15 to 30 minutes. The cold will slow the insect down, making it easier to observe and draw. When you are done drawing it, release it back where you originally found it.

Reminder: Get Locusts: Insects on the Move *and* The Fiddlehoppers: Crickets, Katydids, and Locusts *for lesson 8. You may also be reading passages from* Social Life in the Insect World, *the Bible, and* On the Banks of Plum Creek.

Lesson 8

Locusts

Read *Jack's Insects*, chapter 8, "All Full of Glory and Grandeur."

Narration

1. Describe the eating habits of the Plague Locust.

2. Explain how Signor Matthei saved Cyprus from the locusts. Draw a diagram if you would like to.

In this book from 1917, reference is made to the wall of calico and leather described in *Jack's Insects*:

> "The only successful method of exterminating locusts was adopted in Cyprus in 1881. Since 1609 the island had been a wilderness. Matthei, conversant with the habits of the larvae, erected an insurmountable wall of calico and leather round the main area. Unable to pass the smooth leather, the locusts fell into the trench dug beneath" (James Hastings, Encyclopedia of Religion and Ethics, Part 15, p. 126).

3. Tell about the egg laying of Australian Locust.

_____
_____
_____
_____
_____
_____
_____

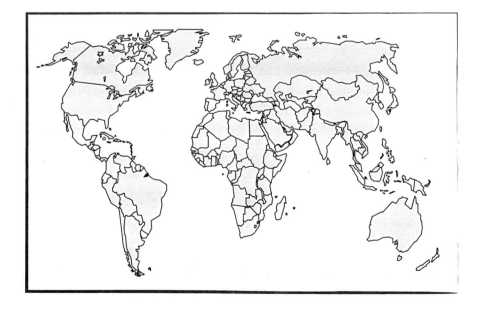

On this world map, mark and label the region where the locust would be known best from writings. Also mark and label the countries that the Plague Locust mentioned in his history of principal locust campaigns. Last, mark and label the continent where Jack and Maggie met another locust.

Find Out More

"The importance of insects to us both as sources of good or evil, I shall endeavour to prove at large hereafter; but for the present, taking this for granted, it necessarily follows that the study of them must also be important. For when we suffer from them, if we do not know the cause, how are we to apply a remedy that may diminish or prevent their ravages? Ignorance in this respect often occasions us to mistake our enemies for our friends, and our friends for our enemies; so that when we think to do good we only do harm, destroying the innocent and letting the guilty escape" (William Kirby, An Introduction to Entomology, letter II).

Read *Locusts: Insects on the Move* by Sandra Markle and *The Fiddlehoppers: Crickets, Katydids, and Locusts* by Phyllis Perry to learn more about locusts and how they interact with their environment.

If you are interested in learning even more about locusts, read chapter 20, "The Grey Locust," in *Social Life in the Insect World* by Jean Henri Fabre.

Locust plagues were miserable to live through. We are given a brief explanation of a locust plague in Exodus chapter 10 in the Bible. God used a locust plague to help convince the pharaoh of Egypt to let the Jews go. Laura Ingalls Wilder gives us a more detailed account of what it is like to live through an infestation of locusts. Read chapters 25 and 26 in Laura Ingalls Wilder's book, *On the Banks of Plum Creek*.

The nineteenth-century entomologist, William Kirby, wrote in his book, *An Introduction to Entomology*, about several locust infestations in Europe and the damage the locusts caused.

"From 1778 to 1780 the empire of Marocco was terribly devastated by them; every green thing was eaten up, not even the bitter bark of the orange and pomegranate escaping – a most dreadful famine ensued. The poor were seen to wander over the country deriving a miserable subsistence from the roots of plants; and women and children followed the camels from whose dung they picked the undigested grains of barley, which they devoured with avidity: in consequence of this, vast numbers perished, and the roads and streets exhibited the unburied carcasses of the dead" (William Kirby, *An Introduction to Entomology*, letter VII).

"The neighbouring kingdom of Spain has often suffered from the ravages of locusts. So recently as May, 1841, an article in the *Constitutionnel* French newspaper states as follows: 'Such immense quantities of locusts have appeared this year in Spain that they threaten in some places entirely to destroy the crops. At Daimiel, in the province of Ciudad-Real, three hundred persons are constantly employed in collecting these destructive insects, and though they destroy seventy or eighty sacks every day, they do not appear to diminish. There is something frightful in the appearance of these locusts proceeding in divisions, some of which are a league in length and 2000 paces in breadth. It is sufficient if these terrible columns stop half an hour on any spot, for every thing growing on it – vines, olive-trees, and corn – to be entirely destroyed. After they have passed, nothing remains but the large branches and the roots, which, being under ground, have escaped their voracity' " (William Kirby, *An Introduction to Entomology*, letter VII).

"And in a late work of travels in the same country we find the following passage: 'During our ride (from Cordova to Seville), we observed a number of men advancing in skirmishing order across the country, and thrashing the ground most savagely with long flails. Curious to know what could be the motives for this Xerxes-like treatment of the earth, we turned out of the road to inspect their operations, and found they were driving a swarm of locusts into a wide piece of linen, spread on the ground some distance before them, wherein they were made prisoners. These animals are about three times the size of an English grasshopper. They migrate from Africa, and their spring visits are very destructive; for in a single night they will entirely eat up a field of corn' " (William Kirby, *An Introduction to Entomology*, letter VII).

The following websites have good pictures of locusts laying eggs, locust eggs, and locust nymphs.

http://simplycm.com/jacks0801
http://simplycm.com/jacks0802

Always ask your parent's permission before visiting any website.

Here's what I found out . . .

_____

_____

_____

Nature Study

When you are outside doing nature study, look for hopping insects like grasshoppers and locusts.

On pages 94 and 95 you can draw the hopping insects that you see. If you don't see them live, you can draw them by looking at a photograph. Be sure to label each drawing.

Lesson 9

Katydid

Read *Jack's Insects*, chapter 9, "A Very Distinguished Musician."

Narration

1. Explain the confusing scientific classification of the katydid.

2. Tell about the physical differences that the katydid and Jack and Maggie discovered between themselves, as well as the music-making differences.

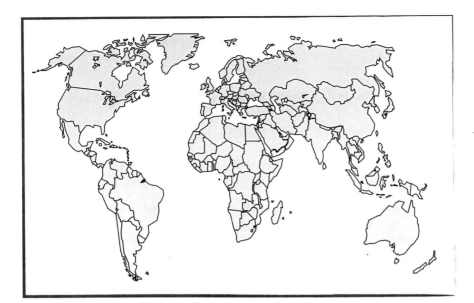

On this world map, mark and label the place where Jack and Maggie found the katydid.

"Why were insects made attractive, if not, as Ray well expresses it, that they might ornament the universe and be delightful objects of contemplation to man?" (William Kirby, An Introduction to Entomology, letter II).

Nature Study

When you are outside doing nature study, look for katydids and other hopping insects.

On pages 94 and 95 you can draw the hopping insects that you see. Try to draw each insect at least its actual size. If your drawing is larger than the insect itself, add a less-detailed sketch of it actual size or a note telling its actual size.

Reminder: Get Grasshoppers and Crickets of North America for lesson 10.

Lesson 10

Katydid Concert

Read *Jack's Insects*, chapter 10, "Jack and Maggie Go To a Concert."

Narration

1. Describe in full the concert the katydids gave for Jack and Maggie.

___

2. Explain the duty and secret of the katydids.

___

Find Out More

Read *Grasshoppers and Crickets of North America* by Sara Swan Miller.

Here's what I found out . . .

_____
_____
_____
_____
_____
_____
_____
_____
_____
_____

Nature Study

When you are outside doing nature study, look for katydids and other hopping insects. On pages 94 and 95 you can draw the hopping insects that you see. You will be most likely to hear a katydid concert if you go outdoors on a summer night. Listen for the sounds of insects to help you know they are around and to locate them. Katydids don't move around a lot; if you hear one singing, you will probably be able to follow the sound and shine a flashlight on it. The light may startle it at first, but sometimes it will get used to the light and start singing again; then you can watch it perform its concert.

Reminder: Use the Exam Questions on page 83 to review what you have learned so far.

Lesson 11

Cicada

Read *Jack's Insects*, chapter 11, "A Very Classical Insect."

Narration
Tell what the ancient Greeks wrote about cicadas and some of their practices including cicadas.

_____
_____
_____
_____
_____
_____
_____

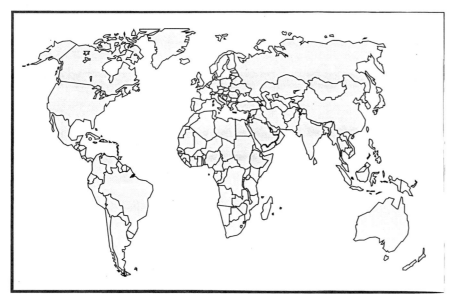

The cicada Jack and Maggie talked with was in the same country as the katydid. Mark and label that country on this map.

"Captain Hancock informs me that the Brazilian Cicada sing so loud as to be heard at the distance of a mile. This is as if a man of ordinary stature, supposing his powers of voice increased in the ratio of his size, could be heard all over the world" (William Kirby, An Introduction to Entomology, letter XXIV).

Nature Study

When you are outside doing nature study, look for cicadas. Keep track of which insects you have seen by checking them off the list on pages 86 through 90.

On pages 96 and 97 you can draw the cicadas that you see. If you don't see them live, you can draw them by looking at a photograph.

Reminder: Get Cicadas!: Strange and Wonderful for lesson 12. You will also be reading from Social Life in the Insect World.

Lesson 12

Cicada Development

Read *Jack's Insects*, chapter 12, "A Remarkable Autobiography."

Narration

1. Starting with the egg hatching, describe the development of the cicada.

_____
_____
_____
_____
_____

2. Tell all you know about how a cicada eats in its "imperfect" stage.

_____
_____
_____
_____
_____

3. Explain how the cicada eats in its "perfect" stage.

_____
_____
_____
_____

Find Out More

Read Cicadas!: Strange and Wonderful by Laurence Pringle and Social Life in the Insect World by Jean Henri Fabre, chapters 1 to 4, "The Fable of the Cigale and the Ant," "The Cigale Leaves Its Burrow," "The Song of the Cigale," and "The Cigale – The Eggs and Their Hatching."

To see some pictures of a cicada molting, visit http://simplycm.com/jacks1201. Remember to ask your parent's permission before visiting any website.

Here's what I found out . . .

_____
_____
_____
_____
_____
_____
_____
_____

Nature Study

When you are outside doing nature study, look for cicadas.

On pages 96 and 97 you can draw the cicadas that you see. You will probably be able to hear them easier than you can see them; they are masters of camouflage and blend well with the trees. Look for their molted skin on the trees and on the ground around the trees.

 _____

Reminder: You will need your insect field guide and Animal Lives: Bees and Wasps (Qeb Animal Lives) for lesson 13.

Lesson 13

Great Golden Digger Wasp

Read *Jack's Insects*, chapter 13, "Jack and Maggie Are Found To Be Not Quite In Tune."

Narration

1. Describe how the Great Golden Digger Wasp transported the cicada to her nursery larder.

_____
_____
_____
_____

2. Tell all you know about the nursery larder.

_____
_____
_____
_____
_____

3. Define "ovipositing."

_____
_____
_____

4. Explain the purpose the cicada served for the wasp.

_____
_____
_____
_____

5. Tell about the life cycle of the Great Golden Digger Wasp.

_____
_____
_____
_____

Find Out More

Look up information on Great Golden Digger Wasps and Cicada Killer Wasps in a field guide. Which wasp fits the information given in *Jack's Insects* the best? Give your reasons and cite examples from *Jack's Insects* and from your research sources.

Read *Animal Lives: Bees and Wasps* (Qeb Animal Lives) by Sally Morgan.

Here's what I found out . . .

_____
_____
_____
_____
_____
_____

_____
_____
_____
_____
_____
_____
_____
_____
_____
_____
_____
_____
_____
_____

Nature Study

When you are outside doing nature study, look for cicadas and wasps.

On pages 96 and 97 you can draw the cicadas and wasps that you see. Be sure to observe wasps at a safe distance to avoid getting stung. If you are relaxed and quiet, you can get closer than if you are nervous; but remember to keep a safe distance and don't try to touch them. Anxious people agitate wasps.

Reminder: Get The Glow-Worm and Other Beetles for lesson 14. You should be able to find it available free at http://simplycm.com/bf/98101.

Lesson 14

Candle Fly and Lantern Fly

Read *Jack's Insects*, chapter 14, "A Queer Trial Without Any Verdict."

Narration

1. Describe the Candle Fly.

_____
_____
_____
_____

2. Describe the Great Lantern Fly.

_____
_____
_____
_____

3. Tell about the trial with the Lantern Fly and the Candle Fly.

_____
_____
_____
_____
_____

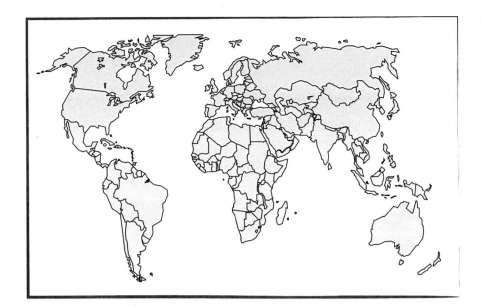

On this world map, mark and label the three locations where Jack and Maggie talked with luminous insects.

Find Out More

Read chapter 1, "The Glow-Worm," in *The Glow-Worm and Other Beetles* by Jean Henri Fabre.

William Kirby wrote at length about luminous insects in his *An Introduction to Entomology*.

"If you take one of these glow-worms home with you for examination, you will find that in shape it somewhat resembles a caterpillar, only that it is much more depressed; and you will observe that the light proceeds from a pale-coloured patch that terminates the under side of the abdomen. It is not, however, the larva of an insect, but the perfect female of a winged beetle, from which it is altogether so different that nothing but actual observation could have inferred the fact of their being the sexes of the same insect" (William Kirby, *An Introduction to Entomology*, letter XXV).

"It has been supposed by many that the males of the different species of *Lampyris* do not possess the property of giving out any light; but it is now ascertained that this supposition is inaccurate, though their light is much less vivid than that of the female" (William Kirby, *An Introduction to Entomology*, Letter XXV).

"Two of the most conspicuous of this tribe are the *F. laternaria* and *F. candelaria;* the former a native of South America, the latter of China. Both, as indeed is the case with the whole genus, are supposed to have the material which diffuses their light included in a subtransparent projection of the head. In *F. candelaria* this projection is of a subcylindrical shape, recurved at the apex, above an inch in length, and the thickness of a small quill. In *F. laternaria*, which is an insect two or three inches long, the snout is much larger and broader, and more of an oval shape, and sheds a light the brilliancy of which is said to transcend that of any other luminous insect" (William Kirby, *An Introduction to Entomology*, letter XXV).

"Madame Merian informs us, that the first discovery which she made of this property caused her no small alarm. The Indians had brought her several of these insects, which by daylight exhibited no extraordinary appearance, and she inclosed them in a box until she should have an opportunity of drawing them, placing it upon a table in her lodging-room. In the middle of the night the confined insects made such a noise as to awake her, and she opened the box, the inside of which to her great astonishment appeared all in a blaze; and in her fright letting it fall, she was not less surprised to see each of the insects apparently on fire. She soon, however, divined the cause of this unexpected phenomenon, and re-inclosed her brilliant guests in their place of confinement. She adds, that the light of one of these *Fulgora* is sufficiently bright to read a newspaper by; and though the tale of her having drawn one of these insects by its own light is without foundation, she doubtless might have done so if she had chosen" (William Kirby, *An Introduction to Entomology*, letter XXV).

Here's what I found out . . .

_____
_____
_____
_____
_____
_____
_____
_____
_____
_____

Nature Study

When you are outside doing nature study, look for luminous insects. Most likely the only ones you will find will be fireflies. Check them off the list on page 89 if you see them.

On page 98 you can draw the luminous insects that you see. Fireflies usually start flashing at twilight and will flash for several hours.

Lesson 15

Cucujo

Read *Jack's Insects*, chapter 15, "A Flying Visit to Mexico."

Narration

1. Describe the Cucujo.

_____
_____
_____
_____
_____
_____
_____

2. Tell about the use of the Cucujo in Mexico.

_____
_____
_____
_____
_____
_____
_____
_____
_____
_____

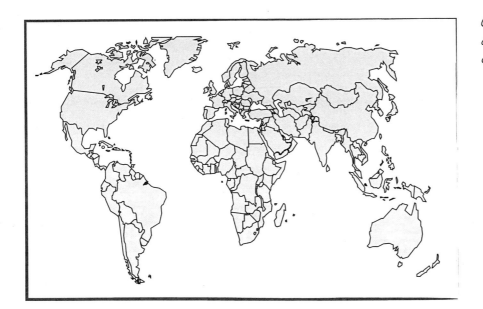

On this world map, mark and label Mexico, the home of the cucujo.

Find Out More

Here is some more information about the cucujo from Kirby's *An Introduction to Entomology*.

"The light emitted by the two thoracic tubercles alone is so considerable, that the smallest print may be read by moving one of these insects along the lines; and in the West India Islands, particularly in St. Domingo, where they are very common, the natives were formerly accustomed to employ these living lamps, which they call *Cucuij*, instead of candles in performing their evening household occupations. In travelling at night, they used to tie one to each great toe; and in fishing and hunting required no other flambeau. Southey has happily introduced this insect in his '*Madoc*,' as furnishing the lamp by which Coatel rescued the British hero from the hands of the Mexican priests" (William Kirby, *An Introduction to Entomology*, letter XXV).

"These insects are also applied to purposes of decoration. On certain festival days, in the month of June, they are collected in great numbers, and tied all over the garments of the young people, who gallop through the streets on horses similarly ornamented, producing on a dark evening the effect of a large moving body of light. On such occasions the lover displays his gallantry by decking his mistress with these living gems. And according to P. Martire, 'many wanton wilde fellowes' rub their faces with the flesh of a killed Cucuius, as boys with us use phosphorus, 'with purpose to meet their neighbours with a flaming countenance,' and derive amusement from their fright" (William Kirby, *An Introduction to Entomology*, letter XXV).

"If we are to believe Mouffet (and the story is not incredible), the appearance of the tropical fire-flies on one occasion led to a more important result than might have been expected from such a cause. He tells us, that when Sir Thomas Cavendish and Sir Robert Dudley first landed in the West Indies, and saw in the evening an infinite number of moving lights in the woods, which were merely these insects, they supposed that the Spaniards were advancing upon them, and immediately betook themselves to their ships: – a result as well entitling the Elaters to a commemoration feast as a similar good office the land-crabs of Hispaniola, which, as the Spaniards tell (and the story is confirmed by an anniversary *Fiesta de los Cangrejos*), by their clattering – mistaken by the enemy for the sound of Spanish cavalry close upon their heels – in like manner scared away a body of English invaders of the city of St. Domingo" (William Kirby, *An Introduction to Entomology*, letter XXV).

Here's what I found out . . .

Nature Study

When you are outside doing nature study, look for luminous insects.

On page 98 you can draw the luminous insects that you see. If you don't see them live, you can draw them by looking at a photograph. If it is spring, you may be able to find firefly grubs by digging in the ground. Those grubs are luminous and will glow, even in the daylight.

Reminder: Get the book Fireflies (Bugs, Bugs, Bugs) for lesson 16.

Lesson 16

Fireflies

Read *Jack's Insects*, chapter 16, "Jack and Maggie Drop Off with the Conversation."

Narration

1. Tell about the flight on the fireflies.

_____
_____
_____
_____
_____
_____
_____

2. Do you agree or disagree with the fireflies regarding the scientist and his experiment? Explain why.

_____
_____
_____
_____
_____
_____
_____
_____

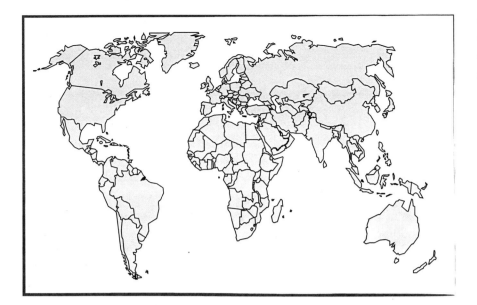

On this world map, mark and label the country where Jack and Maggie talked with the fireflies.

Find Out More

Read *Fireflies* (Bugs Bugs Bugs) by Margaret Hall.

Here's what I found out . . .

_____
_____
_____
_____
_____
_____
_____
_____
_____
_____
_____
_____

_____

_____

_____

Nature Study

When you are outside doing nature study, look for luminous insects.

On page 98 you can draw the luminous insects that you see. Many children find it almost irresistible to collect fireflies in a glass jar. Unfortunately, many fireflies die that way. Here's how to keep them alive. Put screen or netting over the top of the jar, rather than a solid lid. Place a small square of damp sponge in the jar; that's all the fireflies need to stay alive overnight. Release them in the morning.

 _____

Reminder: Get the book Spiders for lesson 17.

Lesson 17

Diving-Bell Spider

Read *Jack's Insects*, chapter 17, "A Visit to Diving-Bell Hall."

Narration

1. Tell what happened to Jack and Maggie after the fireflies had gone out.

_____
_____
_____
_____

2. Describe how the water-spider made Diving-bell Hall.

_____
_____
_____
_____
_____
_____
_____

3. Explain what the water-spider eats and how she catches her prey.

_____
_____
_____
_____
_____

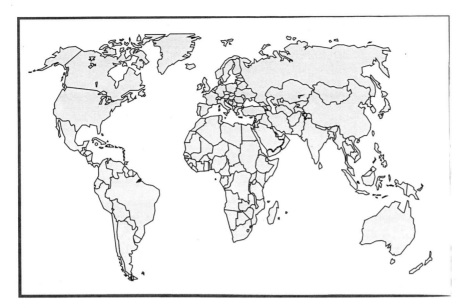

On this world map, mark and locate the country where Jack and Maggie encountered the diving-bell spider.

Find Out More

Read *Spiders* by Seymour Simon.

Read William Kirby's observations of the diving-bell spider.

"The habitation of *Argyroneta aquatica*, the other spider to which I alluded, is chiefly remarkable for the element in which it is constructed and the materials that compose it. It is built in the midst of water, and formed, in fact, of air! Spiders are usually terrestrial, but this is aquatic, or rather amphibious; for though she resides in the midst of water, in which she swims with great celerity, sometimes on her belly, but more frequently on her back, and is an admirable diver, she not unfrequently hunts on shore, and having caught her prey, plunges with it to the bottom of the water. Here it is she forms her singular and unique abode. She would evidently have but a very uncomfortable time were she constantly wet, but this she is sagacious enough to avoid; and by availing herself of some well-known philosophical principles, she constructs for herself an apartment in which, like the mermaids and sea-nymphs of fable, she resides in comfort and security. The following is her process. First she spins loose threads in various directions attached to the leaves of aquatic plants, which may be called the frame-work of her chamber, and over them she spreads a transparent varnish resembling liquid glass, which issues from the middle of her spinners, and which is so elastic that it is capable of great expansion and contraction; and if a hole be made in it, it immediately closes again. Next she spreads over her belly a pellicle of the same material, and ascends to the surface. The precise mode in which she transfers a bubble of air beneath this pellicle is not accurately known; but from an observation made by the ingenious author of the little work from which this account is abstracted, he concludes that she draws the air into her body by the anus, which she presents to the surface of the pool, and then pumps it out from an opening at the base of the belly between the pellicle and that part of the body, the hairs of which keep it extended. Clothed with this aerial mantle, which to the spectator seems formed of resplendent quicksilver, she plunges to the bottom, and, with as much dexterity as a chemist transfers gas with a gas-holder, introduces her bubble of air beneath the roof prepared for its reception. This manoeuvre she repeats ten or twelve times, until at length in about a quarter of an hour she has transported as much air as suffices to expand her apartment to its intended extent, and now finds herself in possession of a little aerial edifice, I had almost said an enchanted palace, affording her a commodious and dry retreat in the very midst of the water. Here she reposes unmoved by the storms that agitate the surface of the pool, and devours her prey at ease and in safety. Both sexes form these lodgings. At a particular season of the year the male quits his apartment, approaches that of the female, enters it, and enlarging it by the bubble of air that he carries with him, it becomes a common abode for the happy pair. The spider which forms these singular habitations is one of the largest European species, and in some countries not uncommon in stagnant pools" (William Kirby, *An Introduction to Entomology*, letter XIV).

To see pictures and videos of diving-bell spiders visit http://simplycm.com/jacks1701. Remember to ask your parent's permission before visiting any website.

"We imagine that nothing short of human intellect can be equal to the construction of a diving-bell or an air-pump—yet a spider is in the daily habit of using the one, and, what is more, one exactly similar in principle to ours, but more ingeniously contrived; by means of which she resides unwetted *in the bosom of the water, and procures the necessary supplies of air by a much more simple process than our alternating buckets"* (William Kirby, An Introduction to Entomology, letter 1).

Here's what I found out . . .

_____
_____
_____
_____
_____
_____
_____
_____
_____
_____
_____
_____
_____
_____
_____
_____

Nature Study

When you are outside doing nature study, look for spiders.

On pages 99 and 100 you can draw the spiders that you see. If you don't see them live, you can draw them by looking at a photograph. You might want to use the refrigerator trick to be able to look more closely at the spiders.

Lesson 18

Raft Spider

Read *Jack's Insects*, chapter 18, "The Raft-Spider Does Her Very Best."

Narration

1. Describe the raft-spider's raft.

_____
_____
_____
_____
_____
_____
_____
_____

2. Tell about the raft ride across the water.

_____
_____
_____
_____
_____
_____
_____
_____
_____
_____

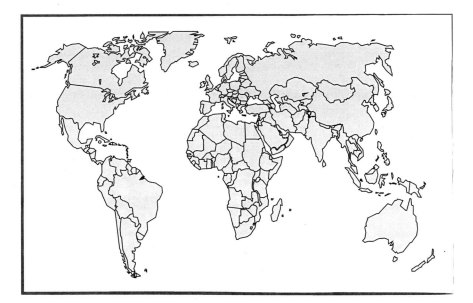

The raft spider was in the same country as the diving-bell spider. Mark and label that country on this map.

Find Out More

Here is a description of one man's observations of a raft spider:

"The Rev. R. Sheppard has often noticed, in the fen ditches of Norfolk, a very large spider, which actually forms a *raft* for the purpose of obtaining its prey with more facility. Keeping its station upon a ball of weeds about three inches in diameter, probably held together by slight silken cords, it is wafted along the surface of the water upon this floating island, which it quits the moment it sees a drowning insect, – not, as you may conceive, for the sake of applying to it the process of the Humane Society, but of hastening its exit by a more speedy engine of destruction. The booty thus seized it devours at leisure upon its raft, under which it retires when alarmed by any danger" (William Kirby, *An Introduction to Entomology*, letter XIII).

With your parent's permission, watch this video on raft spiders. Be sure to pause the video at the end so you can read more information about raft spiders.

http://simplycm.com/jacks1801

Here's what I found out . . .

_____
_____
_____
_____
_____
_____
_____
_____
_____
_____
_____
_____
_____
_____

Nature Study

When you are outside doing nature study, look for spiders. Keep track of which ones you have seen by checking them off the list on page 90.

On pages 99 and 100 you can draw the spiders that you see. Be sure to label each drawing.

Lesson 19

Swarming Bees

Read *Jack's Insects*, chapter 19, "Jack and Maggie Are Presented At Court."

Narration

Tell about Jack's and Maggie's experience in the Queen Bee's court.

_____
_____
_____
_____
_____
_____
_____
_____
_____

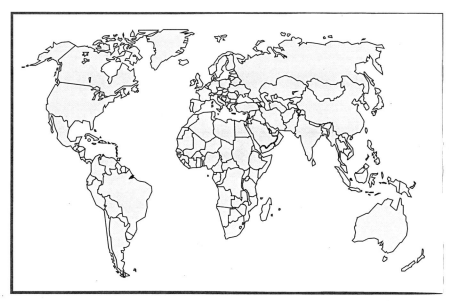

Jack and Maggie stay in the same country to learn about bees. Mark and label where they are in this chapter.

Nature Study

When you are outside doing nature study, look for bees. Keep track of which ones you have seen by checking them off the list on page 90.

On pages 101 and 102 you can draw the bees that you see. If you don't see them live, you can draw them by looking at a photograph. Bees are not as aggressive as some wasps and hornets can be. So you can usually get fairly close as long as you remain calm and quiet. But remember to keep a safe distance and don't touch the bees.

Reminder: Get Bramble-Bees and Others for lesson 20. Bramble-Bees and Others is available free at http://simplycm.com/bf/98102.

Lesson 20

Unusual Bees and the Honey-pot Ant

Read *Jack's Insects*, chapter 19, "Into the Bee-Hive And Out of the Book."

Narration

1. Describe the carding of the Carding Bee.

_____
_____
_____
_____
_____
_____
_____

2. Tell about the Leaf-cutting bee's cradle.

_____
_____
_____
_____
_____
_____
_____

3. Tell about the Honey-pot ant.

_____

_____

_____

_____

_____

_____

_____

4. Explain how Jack and Maggie got out of the book.

_____

_____

_____

_____

_____

_____

_____

_____

Find Out More

Read chapter 8, "The Leaf-Cutters," and chapter 9, "The Cotton-Bees," from *Bramble-Bees and Others* by Jean Henri Fabre. The cotton bees he refers to are a type of carding bee.

Read Kirby's observations on the Carding bee.

"*Bombus Muscorum*, and some other species of humble-bees, cover their nests with a roof of moss. M. P. Huber having placed a nest of the former under a bell-glass, he stuffed the interstices between its bottom and the irregular surface on which it rested with a linen cloth. This cloth, the bees, finding themselves in a situation where no moss was to be had, tore thread from thread, carded it with their feet into a felted mass, and applied it to the same purpose as moss, for which it was nearly as well adapted. Some other humble-bees tore the cover of a book with which he had closed the top of the box that contained them, and made use of the detached morsels in covering their nest" (William Kirby, *An Introduction to Entomology*, letter XXVII).

Read William Kirby's observations on Leaf-cutting bees.

"The mother-bee first excavates a cylindrical hole eight or ten inches long, in a horizontal direction, either in the ground or in the trunk of a rotten willow-tree, or occasionally in other decaying wood. This cavity she fills with six or seven cells wholly composed of portions of leaf, of the shape of a thimble, the convex end of one closely fitting into the open end of another. Her first process is to form the exterior coating, which is composed of three or four pieces of larger dimensions than the rest, and of an oval form. The second coating is formed of portions of equal size, narrow at one end, but gradually widening towards the other, where the width equals half the length. One side of these pieces is the serrate margin of the leaf from which it was taken, which, as the pieces are made to lap one over the other, is kept on the outside, and that which has been cut within. The little animal now forms a third coating of similar materials, the middle of which, as the most skilful workman would do in similar circumstances, she places over the margins of those that form the first tube, thus covering and strengthening the junctures. Repeating the same process, she gives a fourth and sometimes a fifth coating to her nest, taking care, at the closed end or narrow extremity of the cell, to bend the leaves so as to form a convex termination. Having thus finished a cell, her next business is to fill it to within half a line of the orifice with a rose-coloured conserve composed of honey and pollen, usually collected from the flowers of thistles; and then having deposited her egg, she closes the orifice with three pieces of leaf so exactly circular, that a pair of compasses could not define their margin with more truth; and coinciding so precisely with the walls of the cell, as to be retained in their situation merely by the nicety of their adaptation. After this covering is fitted in, there remains still a concavity which receives the convex end of the succeeding cell; and in this manner the indefatigable little animal proceeds until she has completed the six or seven cells which compose her cylinder" (William Kirby, *An Introduction to Entomology*, letter XIV).

"The process which one of these bees employs in cutting the pieces of leaf that compose her nest is worthy of attention. Nothing can be more expeditious: she is not longer about it than we should be with a pair of scissors. After hovering for some moments over a rose-bush, as if to reconnoitre the ground, the bee alights upon the leaf which she has selected, usually taking her station upon its edge, so that the margin passes between her legs. With her strong mandibles she cuts without intermission in a curve line, so as to detach a triangular portion. When this hangs by the last fibre, lest its weight should carry her to the ground, she balances her little wings for flight, and the very moment it parts from the leaf flies off with it in triumph; the detached portion remaining bent between her legs in a direction perpendicular to her body. Thus, without rule or compasses do these diminutive creatures mete out the materials of their work into portions of an ellipse, into ovals or circles, accurately accommodating the dimensions of the several pieces of each figure to each other. What other architect could carry impressed upon the tablet of his memory the entire idea of the edifice which he has to erect, and, destitute of square or plumb line, cut out his materials in their exact dimensions without making a single mistake? Yet this is what our little bee invariably does. So far are human art and reason excelled by the teaching of the Almighty" (William Kirby, *An Introduction to Entomology*, letter XIV).

Here's what I found out . . .

_____
_____
_____
_____
_____
_____
_____
_____
_____
_____
_____
_____
_____

_____

_____

_____

_____

_____

_____

_____

_____

_____

_____

Nature Study

When you are outside doing nature study, look for bees and ants. Keep track of which ones you have seen by checking them off the list on page 90.

On pages 101 and 102 you can draw the bees and ants that you see. Remember that these drawings are not just art exercises. You are drawing the insects to help you observe, learn, and fully appreciate how God made them.

 _____

Reminder: Use the Exam Questions on page 83 to review what you have learned.

We hope that you've enjoyed this study of insects. Just because you have finished this course doesn't mean that your study of insects has to be over. We hope that you will keep studying them.

> "To follow only the insects that frequent your own garden, from their first to their last state, and to trace all their proceedings, would supply an interesting amusement for the remainder of your life" (William Kirby, An Introduction to Entomology, letter I).

> "For more than twenty years my attention has been directed to them, and during most of my summer walks my eyes have been employed in observing their ways; yet I can say with truth, that so far from having exhausted the subject, within the last six months I have witnessed more interesting facts respecting their history than in many preceding years" (William Kirby, An Introduction to Entomology, letter I).

Exam Questions

Use these questions to help review and evaluate how much you've learned through your reading and observations.

After lesson 10

Narration Questions
1. Describe at least two insects that use camouflage or mimicry to defend themselves.
2. Tell all you know about locusts and the part they have played in mankind's history.

Nature Study Questions
1. What have you noticed (yourself) about a butterfly or caterpillar?
2. Select an insect you have studied in the first half of the book, sketch it and tell all you know about it.

After lesson 20

Narration Questions
1. Tell all you know about the cicada: its history, its development, and its life.
2. Describe at least two luminous insects, telling where they can be found and all you know about them.

Nature Study Questions
1. What have you noticed (yourself) about a spider?
2. Select an insect you have studied in the second half of the book, sketch it and tell all you know about it.

Final Question

William Kirby wrote: "Who then shall dare maintain, unless he has the hardihood to deny that God created them, that the study of insects and their ways is trifling or unprofitable? Were they not arrayed in all their beauty, and surrounded with all their wonders, and made so instrumental . . . to our welfare, that we might glorify and praise him for them?" (*An Introduction to Entomology*, letter II).

What have you learned or observed in your study of insects that supports this statement? Has your study encouraged you to glorify and praise God? How?

Who Was William Kirby?

Sprinkled throughout this study you will find many comments from William Kirby. Kirby was a man who loved insects; he especially loved to study the insects of his home country, England. Entomology was a passion of his during his whole lifetime, from 1759 to 1850. Take a moment to think of other people you may have read about who were alive during that time. You might check your Book of Centuries.

In 1833 he helped to found the Entomological Society of London and became its president for life. He even gave his personal insect collection to the Society, a collection that he had worked on for more than 40 years.

Kirby was also a man who loved the Lord. He was a clergyman, or pastor, in Suffolk. As he studied insects, he wrote several books and papers detailing his observations. But his overall goal was to show how nature, and insects in particular, point people to God, the loving and good Creator: "The author of Scripture is also the author of Nature: and this visible world, by types indeed, and by symbols, declares the same truths as the Bible does by words. To make the naturalist a religious man – to turn his attention to the glory of God, that he may declare his works, and in the study of his creatures may see the loving-kindness of the Lord – may this in some measure be the fruit of my work" (Correspondence, 1800).

Master Insect & Spider List

Check off each insect or spider you find in your nature studies outdoors. If you find one that is not on the list, add it.

Butterflies
- ☐ Monarch
- ☐ Tiger Swallowtail
- ☐ Spicebush Swallowtail
- ☐ Black Swallowtail
- ☐ Anise Swallowtail
- ☐ Giant Swallowtail
- ☐ Cabbage White
- ☐ Common Sulfur
- ☐ Orange Sulfur
- ☐ Common Checkered Skipper
- ☐ Common Roadside Skipper
- ☐ American Copper
- ☐ Lustrous Copper
- ☐ Banded Hairstreak
- ☐ Eastern Tailed Blue
- ☐ Western Tailed Blue
- ☐ Dotted Blue
- ☐ Spring Azure
- ☐ Red Admiral
- ☐ Painted Lady
- ☐ Mourning Cloak
- ☐ Buckeye

- ☐ Question Mark
- ☐ Comma
- ☐ Red-spotted Purple
- ☐ Common Wood Nymph

- ☐ _____
- ☐ _____
- ☐ _____
- ☐ _____
- ☐ _____
- ☐ _____
- ☐ _____
- ☐ _____

Moths

- ☐ Milkweed Tiger Moth
- ☐ Woolly Bear Caterpillar Moth
- ☐ Hummingbird Moth
- ☐ White-lined Sphinx
- ☐ Artichoke Plume Moth
- ☐ Emerald Moth
- ☐ Luna Moth
- ☐ Polyphemus Moth
- ☐ Io Moth
- ☐ Imperial Moth
- ☐ Cecropia Moth

- ☐ _____
- ☐ _____
- ☐ _____
- ☐ _____
- ☐ _____

Caterpillars

- ☐ Monarch
- ☐ Question Mark
- ☐ Mourning Cloak
- ☐ Orange Dog
- ☐ Eastern Tiger Swallowtail
- ☐ Black Swallowtail
- ☐ Cabbage White
- ☐ Comma
- ☐ Woolly Bear
- ☐ Inchworm
- ☐ Imperial Moth
- ☐ Io Moth
- ☐ Luna Moth
- ☐ Cecropia Moth
- ☐ Tomato/Tobacco Hornworm
- ☐ _____
- ☐ _____
- ☐ _____
- ☐ _____
- ☐ _____

Insects and Bugs

- ☐ Praying Mantis
- ☐ Walkingstick
- ☐ House Cricket
- ☐ Field Cricket
- ☐ Bird Grasshopper
- ☐ Lubber Grasshopper
- ☐ Tree Cricket
- ☐ Katydid
- ☐ Cicada

- ☐ Firefly
- ☐ Mayfly
- ☐ Squash Bug
- ☐ Four-lined Plant Bug
- ☐ Scarlet-and-green Leafhopper
- ☐ Green Stink Bug
- ☐ Large Milkweed Bug
- ☐ Boxelder Bug
- ☐ Red Milkweed Beetle
- ☐ Ladybug
- ☐ Ground Beetle
- ☐ Stag Beetle
- ☐ Japanese Beetle
- ☐ May Beetle
- ☐ Click Beetle
- ☐ Tiger Beetle
- ☐ Soldier Beetle
- ☐ Green Lacewing
- ☐ Green Darner (Dragonfly)
- ☐ Ebony Jewelwing (Damselfly)
- ☐ Crane Fly
- ☐ Robber Fly
- ☐ Long-legged Fly
- ☐ House Fly
- ☐ Fruit Fly

- ☐ _____
- ☐ _____
- ☐ _____
- ☐ _____
- ☐ _____
- ☐ _____
- ☐ _____

Spiders

- ☐ Daddy-long-legs
- ☐ Jumping Spider
- ☐ Wolf Spider
- ☐ Nursery-web Spider
- ☐ Fishing Spider
- ☐ Crab Spider
- ☐ American House Spider
- ☐ Yellow Garden Spider

- ☐ _____
- ☐ _____
- ☐ _____
- ☐ _____
- ☐ _____

Bees, Wasps, and Ants

- ☐ Bumble Bee
- ☐ Honey Bee
- ☐ Paper Wasp
- ☐ Ichneumon Wasp
- ☐ Great Golden Digger Wasp
- ☐ Cicada Killer
- ☐ Mud Dauber
- ☐ Mason Bees
- ☐ Honey Ant
- ☐ Carpenter Ant
- ☐ Black Ant

- ☐ _____
- ☐ _____

My Drawings of
Butterflies and Moths

My Drawings of
Butterflies and Moths

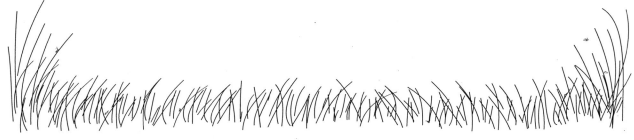

My Drawings of
Caterpillars

My Drawings of
Hopping Insects

My Drawings of
Hopping Insects

My Drawings of
Cicadas and Wasps

My Drawings of
Cicadas and Wasps

My Drawings of
Luminous Insects

My Drawings of
Spiders

My Drawings of
Spiders

My Drawings of
Bees and Ants

My Drawings of
Bees and Ants

My Drawings of
Flies, Beetles, and Bugs

My Drawings of
Flies, Beetles, and Bugs

My Drawings of
Flies, Beetles, and Bugs

My Drawings of
Flies, Beetles, and Bugs